Impact on Earth

The Impact of
FOOD AND FARMING

Cynthia O'Brien

Crabtree Publishing Company
www.crabtreebooks.com

CRABTREE
PUBLISHING COMPANY
WWW.CRABTREEBOOKS.COM

Author: Cynthia O'Brien

Editorial director: Kathy Middleton

Editor: Ellen Rodger

Picture Manager: Sophie Mortimer

Design Manager: Keith Davis

Children's Publisher: Anne O'Daly

Production coordinator and prepress: Ken Wright

Print coordinator: Katherine Berti

Photo credits
(t=top, b= bottom, l=left, r=right, c=center)

Front Cover: All images from Shutterstock

Interior: Aljazerra: Katrina Yu 11; iStock: Marie Art 7, B4lls 25, S Bayram 15, Sera Ficus 5, fishwork 6, JGaunion 14, Kall9 9, Monkey Business Images 23, Anastasia Nurullina 29, HM Proudlove 8, SDI Productions 17, yusnizam 16; Public Domain: Dag Enressen 27; Shutterstock: Benjamin Photography 22, Marius Dobilas 13, Roman Drotyk 4, FenlioQ 10, Yein Jeon 24, Ben Petcharapiracht 26, Ratckova 18, Joshua Resnick 28, smereka 12, BG Smith 19, Sunshine Seeds 21, YuRi Photolife 20.

Library and Archives Canada Cataloguing in Publication

Title: The impact of food and farming / Cynthia O'Brien.
Names: O'Brien, Cynthia (Cynthia J.), author.
Description: Series statement: Impact on Earth | Includes index.
Identifiers: Canadiana (print) 20190236671 |
  Canadiana (ebook) 2019023668X |
  ISBN 9780778774365 (hardcover) |
  ISBN 9780778774624 (softcover) |
  ISBN 9781427125156 (HTML)
Subjects: LCSH: Agriculture—Environmental aspects—Juvenile literature. | LCSH: Food industry and trade—Environmental aspects—Juvenile literature. | LCSH: Food supply—Environmental aspects—Juvenile literature.
Classification: LCC TD195.F57 O27 2020 | DDC J338.1—dc23

Library of Congress Cataloging-in-Publication Data

Names: O'Brien, Cynthia (Cynthia J.), author.
Title: The impact of food and farming / Cynthia O'Brien.
Description: New York, New York : Crabtree Publishing Company, [2020] | Series: Impact on Earth | Includes index.
Identifiers: LCCN 2019053923 (print) | LCCN 2019053924 (ebook)
  ISBN 9780778774365 (hardcover) |
  ISBN 9780778774624 (paperback) |
  ISBN 9781427125156 (ebook)
Subjects: LCSH: Agriculture--Environmental aspects--Juvenile literature. | Agricultural ecology--Juvenile literature. | Nature--Effect of human beings on--Juvenile literature. | Food supply--Juvenile literature. | Food industry and trade--Environmental aspects--Juvenile literature.
Classification: LCC S589.75 O27 2020 (print) |
LCC S589.75 (ebook) | DDC 577.5/5--dc23
LC record available at https://lccn.loc.gov/2019053923
LC ebook record available at https://lccn.loc.gov/2019053924

Published in 2020 by Crabtree Publishing Company

All rights reserved. No part of this publication may be reproduced, stored in a retrieval system or be transmitted in any form or by any means, electronic, mechanical, photocopying, recording, or otherwise, without the prior written permission of the copyright owner.

Copyright © Brown Bear Books Ltd. 2020
Brown Bear Books has made every attempt to contact the copyright holder. If you have any information about omissions, please contact licensing@brownbearbooks.co.uk.

Printed in the U.S.A./022020/CG20200102

Published in Canada
Crabtree Publishing
616 Welland Avenue
St. Catharines, ON
L2M 5V6

Published in the United States
Crabtree Publishing
PMB 59051
350 Fifth Ave, 59th Floor
New York, NY 10118

Contents

Farming and Food Today 4
Why Worry? 8
Climate Change 12
Farming and Wildlife 16
Changing Our Ways 20
Future Developments 24
Your Turn! 28
Glossary 30
Find Out More 31
Index 32

Farming and Food Today

There are more than 570 million farms around the world. These include crop farms, livestock farms, and mixed farms.

Crop farms produce grains, vegetables, and fruit. Other farms raise animals for meat or dairy. Mixed farms grow **crops** and keep **livestock**. Family farms are common around the world. Big companies operate very large farms. Much of the food people buy in the supermarket comes from large farms.

Corn crops cover over 90 million acres (36 million hectares) of the United States. The U.S. is the world's leading corn grower.

TECHNOLOGY SOLUTIONS

Cleaning Up

Fish waste and chemicals from fish farms can pollute nearby waters. Technology is making fish farming more **sustainable**. Filter systems clean and recycle water. They let out little or no **pollution**. Some ocean farms grow a mix of seaweed and shellfish. This provides a natural cleaning system.

75% of the world's farmland is run by family farms

600 Number of types of fish that fish farmers breed

Fish Farms

For thousands of years, people caught fish from oceans, lakes, and rivers. Today, more than half of the fish we eat comes from fish farms, or **aquaculture**. The fish grow in closed tanks or ponds, or in cages in rivers and seas, until they are ready to be sold.

Raspberries from a local farm are delicious and good for you. Buying raspberries in your area helps local farmers and the environment.

Where Does Our Food Come From?

Long ago, people ate food that grew or came from nearby farms. Today, many people buy food at a grocery store. This food comes from all around the world. Countries buy and sell food from each other. This allows us to buy food that does not grow where we live. We can also buy all kinds of food all year round, even food outside our area's growing season.

Environmental Impact

Is it bad for the **environment** to buy **imported** food? Not always. Imported food has to travel in trucks, ships, trains, and planes. This creates a gas called carbon dioxide (CO_2). The rise in carbon dioxide has been a major reason for **climate change**. However, even the way a food is grown and processed locally can create more carbon dioxide than transporting it.

40 times more fuel is used transporting food by air than by sea

WHAT CAN I DO?

Think Local

To help reduce harm to the environment, eat local, **organic** food that is in season. Try growing food yourself in a garden or on a sunny balcony. Visit a farmer's market and meet the people who grew the food.

Why Worry?

Some ways of farming can cause damage to the environment. For example, growing only one type of crop takes nutrients out of the soil.

The use of **pesticides** and **fertilizers** causes pollution. Farms use huge amounts of water and release gases that contribute to climate change. Rain forests have disappeared to make way for farmland. All of these things are problems, and scientists are looking for new ways to farm that will not harm land or water.

Floods ruin farmers' crops. The flood waters can carry diseases that harm farm animals.

WHAT CAN I DO?

Community Garden

Is there a community garden near where you live? This is a garden that people share. Everyone works together to grow good, healthy food. Then they share the food! Many schools also have community gardens. Get involved!

About 80%
Percentage of the world's rain forest that has been cleared, or cut down, for farmland

Food Security

Food security is a term that means having enough nutritious food to eat every day. Not everyone has food security. Farms produce a lot of food, but people do still go hungry if they do not have money to pay. Increasing floods and **droughts** around the world create conditions in which farmers cannot grow food. Less food means more hunger.

Feeding the World

The world's population is getting bigger. By 2050, there will be about 9.8 billion people living on the planet. We have to think about where we can grow food. People have already taken up land with cities, factories, and highways. Not all land is suitable for farming. Farmers cannot grow crops in the ice or in very dry deserts. They cannot raise animals where there is nowhere to graze.

Improving farming methods is a way to reduce harm. For example, rice fields are usually kept flooded. Farmers can improve the soil and grow more rice if they use dry soil and flooding.

EMERGENCY · EMERGENCY ·

Shared Harvest

Shi Yan (left) runs a farm called Shared Harvest near the city of Beijing, China. Shared Harvest produces organic food and supplies it to hundreds of families living in the city. People can rent land and grow their own food. Other people followed Shi Yan's example and started farms. They want everyone to have good food.

51% Percentage of world's food energy provided by corn, rice, and wheat

90% Percentage of world's food energy provided by only 15 of world's 50,000 edible plant types

Staple Foods

Wheat, corn, and rice are staple foods for many people. A staple food means it is a large part of the diet in a certain area. Staples provide basic **nutrients** to give our bodies energy. Growing large crops of just one food is not good for the soil. Growing more kinds of staple foods will help keep the soil and people healthier.

Climate Change

Climate change is a major challenge. Earth's temperature is heating up and affecting the climate worldwide.

Droughts, flooding, and wildfires are increasingly a problem. All of these weather-related events affect the food supply. Farmers cannot grow food if the land is too hot, dry, or wet. New methods of farming are being developed to help slow down climate change so we can grow food in the future.

Livestock farming is a major source of greenhouse gases. Each year, raising farm animals causes 14.5% of the world's greenhouse gases.

TECHNOLOGY SOLUTIONS

Seaweed Diet
Farm animals produce methane, a greenhouse gas. Scientists are trying to reduce the amount of methane produced by livestock. Adding seaweed to their food was found to help decrease their output of gases. Scientists are also finding ways to breed livestock that release less gas from the time they are born.

12,400 miles (20,000 km)
Distance a car must travel to release the same amount of greenhouse gas released by one dairy cow in one year

Greenhouse Gases

Greenhouse gases such as carbon dioxide trap energy from the Sun near Earth. That makes the planet heat up. Human actions, such as taking airplanes, release these gases into Earth's atmosphere. Methane, another greenhouse gas, traps more heat than carbon dioxide.

Flooding, or furrowing, is an irrigation method. It involves flooding a field with a lot of water.

The Water Problem

Water is one of the world's most precious resources. Only a small amount of the world's water is fresh water. The rest is salt water in the oceans and seas. People use fresh water for drinking and in industry. Farms use fresh water for crops and animals. How do we preserve our limited amount of fresh water for the future? For many years, farms have used **irrigation** methods, such as flooding. This uses much more water than farms need. Changing to new methods will help save water. These methods include collecting rainfall and using sprinklers that only water crops when necessary.

Water Pollution

Pesticides are chemicals that keep insects and other pests from destroying crops. Fertilizers are chemicals that help crops to grow. But these chemicals pollute the water and the soil near the crops. Water pollution also comes from livestock waste that enters the water system. In the United States, agriculture is the main cause of pollution in rivers and streams.

70% Percentage of world's fresh water used by farming

TECHNOLOGY SOLUTIONS

Computer Tracking

How do farmers grow crops in places with little rain? Using water from deep underground can help. Still, farmers have to be careful not to use too much water. Scientists use computers to track exactly how much water the crops need. Using a drip irrigation system gets just enough water directly to the roots of plants.

Farming and Wildlife

Wild land is often cleared of trees to make space for farms. This affects the trees, plants, and animals already living there.

People destroy forests every year for farmland. This is called **deforestation**. Forests provide **habitats** for animals and plants, and play an important role in Earth's health. Trees help prevent climate change by absorbing carbon dioxide and greenhouse gases. They also give off oxygen for animals and humans to breathe.

In Indonesia, palm oil farms have taken the place of wild rain forests. Palm oil is used in many foods and other products.

WHAT CAN I DO?

Plant Trees

If you live in a house, ask a parent if you can plant a tree in your backyard. Talk to your teacher about planting trees at your school. Help plant trees in your community. Trees provide shelter for plants and animals, as well as absorb carbon dioxide.

About 27 soccer fields per minute!

How much of world's forests are lost every year

Agroforestry

As the world's population grows, people will need more land to plant crops. One solution may be agroforestry. This is a way of farming without cutting down all the trees. The trees help to keep the soil moist and protect crops from wind. They provide shade for livestock.

Nature at Risk

Farming is one of the biggest threats to wildlife. Modern farming often uses huge fields for crops. These fields remove plants that insects, birds, and other animals depend on for food and shelter. For example, the corn fields in North America have replaced natural **grasslands**. Milkweed plants used to grow on the grassland. Monarch caterpillars need to eat milkweed in order to turn into butterflies. Without the milkweed, the Monarch butterfly is disappearing.

Bees help crops to grow and produce food. Many bees die from pesticides. Many other insects and animals do not get the different plants they need for food.

EMERGENCY · EMERGENCY ·

Bringing Back Wetlands

Wetlands are areas of land covered by water for part or all of the year. Many wetlands have been drained to be used for farming. Scientists have found that wetlands actually help clean soil of the harmful chemicals used in farming. They also provide the farm with water. Ducks Unlimited Canada (DUC) restores and protects wetlands across Canada. Organizations around the world are now working with farmers to restore wetlands.

70 Number out of the top 100 food crops people eat that bees pollinate

50 times more carbon can be stored by wetlands than rain forests

Wetlands

The world's wetlands are home to many plants and animals such as frogs, fish, and water birds. Wetlands are called carbon sinks. These are places that can store a massive amount of CO_2, which helps the environment by keeping it out of the air. More than half the world's wetlands have disappeared since 1900. Many have been drained or burned for farmland.

Changing Our Ways

We can help the Earth by changing how we farm and eat. Many farmers are turning to natural ways to grow their crops.

Growing different kinds of crops can help the soil. These different plants also provide better habitats for wildlife. When farmers rotate their crops, they keep the soil healthy. Rotating crops means changing the places where their plants grow each year. If people ate less meat, farmers would raise fewer cows and more land could be used to grow crops for people.

This farmer is spreading manure on a field. Manure is a natural fertilizer that keeps the soil rich and moist.

EMERGENCY · EMERGENCY ·

Helping Farmers

Self Help Africa is a charity. It works with farmers in Africa to reduce poverty and hunger. The charity helps farmers to grow crops that need less water. It improves the quality of the seeds farmers plant and also the cows farmers raise for milk and other dairy products.

15 lbs (6.8 kg)
Amount of grain needed to produce 1 lb (0.5 kg) of meat

Organic Farming

Many years ago, all farming was organic, which means done without using chemicals on crops or giving drugs to livestock. Today, many people are going back to organic farming. They use **compost** and **manure** as natural fertilizers. They grow "trap" crops. These are crops that attract certain insects so that they leave the main crops alone.

No More Waste!

Every year, about one third of all the food produced in the world is lost or wasted. This means that everything used to produce the food is also wasted. To produce food, we use water, land, and other resources. Food waste happens when people throw away food that is not spoiled. It happens when restaurants throw away food that they do not use. Sometimes food does not even leave the farm. To make things worse, rotting food releases methane.

A vegetable or fruit may not look perfect, but it is still good to eat!

What Can I Do?

Plan Your Shopping

Try planning a week's meals with your family. Figure out the ingredients you will need. Then write a list to take shopping. This way, you buy only enough food to make these meals. If there are leftovers, save them and reuse them on another day.

1.3 billion tons
Amount of food wasted each year

$1 trillion
Cost of this wasted food

Using Waste

Food waste does not have to end up in a landfill. It can be turned into a fuel using a special method. This method heats the waste and breaks it down. This turns the waste into a gas. That gas can be used for electricity or heat. Food waste also makes excellent fertilizer.

Future Developments

What will farms look like in the future? Many farms will use robots and other new technology as well as new growing methods.

Robots will remove weeds and pick fruit. They will help to keep crops and animals healthy by checking them. Scientists are also looking at the way that plants get nutrients from the soil. This will help farmers to use fewer chemicals to raise their crops.

As the world's population grows, we need more food. Vertical farming (see page 25) produces more food per acre than traditional farming.

TECHNOLOGY SOLUTIONS

Sense It

Sensors are small computers that pick up signals. They are planted in the ground and collect information about the soil. Sensors tell the farmer how many nutrients are in the soil, how much water it needs, and when it needs to be watered. New technology can track pests and test the air.

1915 Year the term "vertical farming" was first used

2012 Year world's first commercial vertical farm opened, in Singapore

Skyscraper Farming

Vertical farming is a way of growing layers of crops in an upward direction. It takes up less area than traditional crops on the ground. It uses no soil and very little water. Some companies are building skyscraper farms. These are tall buildings with vertical farms and spaces for people. This way, the food can grow close to where people need it.

Insects are a good source of protein. Farming insects produces fewer greenhouse gases compared to meat production.

Food for the Future

People are thinking of new and exciting ways to make sure we have good, healthy food in the future. Food engineering is part of that. Food engineers look at the way food is processed, or made into other foods. They look at the best ways to produce food and package it. They also help to design machinery that saves energy. Scientists are also developing food plants that need less water and meat that can be grown in laboratories to replace raising animals on farms.

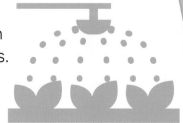

Plant and Insect Diet

Many people already eat foods made from plants. Burgers can be made from pea protein and beet juice and cheese. Many people around the world already eat insects as part of their diet. The future may include more food made from insects, such as flour or pasta. We will also be eating more plants from the sea, such as algae.

2 billion Number of people that eat insects each day

1/3 Fraction that 2 billion people is of world's population

TECHNOLOGY SOLUTIONS

Storing Seeds for the Future

Climate change could wipe out some crops. Seed banks have been set up around the world to hold seeds for growing crops. The seeds are kept safe so people can use them in the future. The biggest seed bank is the Svalbard Global Seed Vault in the Arctic, with more than 980,000 samples.

Your Turn!

Find out how far your food traveled to get to your plate.

Gathering Evidence

Start by collecting information. Keep a food diary for a day. Write down all the meals and food you've eaten, including your packed lunch or school meal. Then find out where each item was produced.

- Look at the food labels on the packages to find which country they came from.
- Some products might have ingredients from different countries.
- You could try doing this with a friend so that you can compare your results.

The food we eat comes from many different countries. Check the labels and write down where your food is from.

Change what you put in your lunch box each season. As well as cutting down food miles, seasonal food is delicious!

Results

Once your survey is complete, print out a world map and mark on it where each food item came from. Figure out how many miles (kilometers) each item traveled to reach your plate. Add them all up to get your total **food miles**. Which food traveled the farthest? Which had the shortest journey? How many were locally produced?

Finding Solutions

Look at the list of foods. Could you have swapped any items for local produce? For the fruits and vegetables, can you think of alternatives that are in season where you live? Rewrite your food diary, with local and seasonal products. Figure out your new total food miles. Have you saved food miles? If so, how many?

Glossary

aquaculture Farming fish and water plants

climate change A change in climate patterns around the world due to the warming of Earth by greenhouse gases

compost Decayed mixture of organic materials

crops Plants grown to be used by people

deforestation Cutting down of large forest areas

droughts Long periods of time with little or no rain

environment Natural world surrounding us

fertilizers Natural or chemical substances used to help plants grow well

food miles The distance food travels before it is sold

grasslands Places where the main plants are grasses

greenhouse gases Gases such as carbon dioxide that build up in the atmosphere and trap heat near Earth

habitats Places where animals and plants live

imported Brought from one country to another

irrigation Supplying with water, for example, by pipes

livestock Farm animals, such as cows, raised for food

manure Solid waste from animals used to improve soil

nutrients Substances that people, animals, and plants need to live and grow

organic Food and animals grown without using chemicals or drugs

pesticides Chemicals used to kill unwanted pests and diseases on plants

pollution The adding of harmful substances into an environment

sustainable Able to be used now and in the future

Find Out More

Books

Bright, Michael. *Field to Plate (Source to Resource)*. Crabtree Publishing, 2017.

Lanz, Helen. *What Shall We Eat?* Sea to Sea Publications, 2012.

Mason, Paul. *Making Our Food Sustainable (Putting the Planet First)*. Crabtree Publishing, 2018.

Websites

This website is all about agriculture, including activities and information for kids.
www.mnagmag.org

Play games and learn about food and farming.
www.myamericanfarm.org

Visit this website to learn about climate change and how farming and deforestation affect it.
www.natgeokids.com/au/discover/geography/general-geography/what-is-climate-change/

Index

C
carbon dioxide 7, 12, 16, 17
chemicals 5, 15, 21, 24, 30
climate change 7, 8, 9, 12, 27
compost 21, 30
corn 4, 11, 18
crop 4, 8, 10, 11, 14, 15, 16, 17, 18, 19, 21, 27

D E F
dairy farming 4, 13, 19
deforestation 16, 30
drought 9, 12
environment 6, 7, 8, 19, 30
fertilizer 8, 15, 23, 30
fish 5, 19
floods 8, 9, 10, 12, 14
food miles 29, 30
food waste 22, 23

G H I
grasslands 18, 30
greenhouse gases 12, 13, 16, 26, 30
habitats 16, 20, 30
irrigation 14, 15, 30

L M N
livestock 4, 12, 15, 17, 21, 30
manure 20, 21, 30
methane 13, 22
nutrients 11, 24, 25, 30

O P
organic farming 21
pesticides 8, 15, 18, 30
pollution 5, 8, 15, 30

R S T
rain forest 8, 9, 19
rice 10, 11
seaweed 5, 13
seed banks 27
staple food 11
Svarlbard Global Seed Vault 27
trap crops 21

V W
vertical farming 25
wetlands 19
wheat 11

3 1333 04991 8384